The transmutation
of metals

The Transmutation of Metals

John Webster
1610-1682

Chapter twenty-nine from

Metallographia: or, an history of metals.

Wherein is declared the signs of ores and minerals both before and after digging, the causes and manner of their generations, their kinds, sorts, and differences; with the description of sundry new Metals, or Semi-Metals, and many other things pertaining to mineral knowledge. As also, the handling and shewing of their Vegetability, and the discussion of the most difficult questions belonging to Mystical Chymistry, as of the Philosophers Gold, their Mercury, the Liquor Alkahest, Aurum potabile, and such like. Gathered forth of the most approved authors that have written in Greek, Latin, or High-Dutch; with some observations and discoveries of the author himself. London, 1671.

Transcribed and the English modernised

by Adam McLean

Alchemy Web Bookshop
2 Craighouse Square
Kilbirnie
KA25 7AF
U.K.

www.alchemywebsite.com

Chapter XXIX

Of the Transmutation of Metals

Concerning this great dispute of the transmuting of one metal into another, we intend not here to demonstrate the possibility of it, for as the Philosopher said well: *Quae experimento oculari videntur, probatione non indigent* [What is seen with the eye in a practical experiment does not need proof]. Only we shall labour to examine, and open the nature of transmutations, and to show some sorts of transmutation, that are common and obvious, and fully as strange as this of metals, and thereby show that it is no such impossible or wonderous thing, as many that would seem wise and learned do labour to make manifest.

So leaving general arguments, as either to confirm or confute, we shall stand upon some instances that may sufficiently demonstrate the way and manner thereof. And as for mutation, transmutation, or that which is commonly called generation, we shall not stand upon the strict logical definition, which we do not perceive, yet is perfectly known, but rather make a search what the nature of it may be; that hereafter it may be better understood and sought into, yet (we suppose) that thus much may be granted, that there can be no generation, but of necessity there must be mutation; for though that all mutations be not generation, yet must every generation, of course, be a mutation, and it is most certain that no mutation can be but by motion ; so that in this the Schools have not far missed it (to give them their due, where, and when they deserve it) that all generation is some kind of mutation, and so must needs be a species of motion: But yet by all this we come not perfectly to know what generation is in its true nature and intrinsical essence and operation, nor the plain and true manner how these mutations are wrought, by the means of motion in or upon matter, but only are left to be puzzled with hard terms, and blind

1

notions, as any person may very well perceive that shall read and seriously consider what these few Authors quoted in the margin have written upon this subject.

We might here fall into those strange mutations that happen in the animal and vegetable kingdom, that appear in the generation of those things; but that would lead us at too far a distance, though they would mightily conduce to open and illustrate the matter we have in hand; and therefore we shall leave them, as things that have been both learnedly and accurately handled by that learned Physician Sennert, Fabritius ab aqua pendente, the incomparable Johannes Marcus Marci, in that curious piece of his styled *Idaea Idaearum operatricium*, by our never-sufficiently-praised countryman Dr Harvey, in his profound piece *De Generatione*, as lastly, by that learned physician and anatomist Dr. Highmore, in his *Treatise Of Generation*, which though little in its bulk, is not little in weight and worth.

But we shall contract ourselves into a closer compass to fall upon some mutations, (or transmutations rather) produced by Nature, or Art, that will sufficiently serve to make forth what we intend concerning the transmutation of metals, only we shall premise this, That all transmutations are made some of these three ways, or by two of them, or all joined together.

1. By adding of something to the thing or subject to be changed that it had not before.

2. Or by taking away, and separating from the subject that is to be changed, that was in it before.

3. Or by reason of motion so to alter, dispose, and order the contexture of the parts, that thereby it appears another thing than what it was before.

And either all of these, or some of them do concur in every transmutation, or else there cannot be any: Now we shall give some instances to make good these particulars, and examine the manner of transmutation by them.

And first of that transmutation which is produced by Nature, in which Art has little or nothing to do, as in petrifying of wood, leaves, moss, grass, and the like, which is for the most part done either by water or other lapidescent juices, or steams turning the forementioned things into a stony matter or substance.

The first we shall name (as being most known unto us) is that famous Dropping Well near the ancient Town of Knaresborough in the West-Riding of Yorkshire, mentioned in our British authors, by our learned Antiquary Mr. Camden, and of late written by Dr. Dean, and Dr. French.

The latter of which says of it thus; If any stick or piece of wood lie in it some weeks, it will be candied over with a stony whitish crust, the inward part of the wood continuing of the same nature as before. But any soft spongy substance, as moss, leaves of trees, &c. into which the water can enter, will thereby in time become seemingly to be of a perfect stony nature and hardness.

Now the cause of this petrifying property, as philosophers call it, is succus lapidescens, i.e. a stony matter, which is in its principiis solutis, for indeed the principia soluta of all things, whether animals, vegetables, metals, or minerals, are in a liquid form, and are concreted by degrees, by a natural heat separating from them all accidental humidities, and fixing them into their proper species.

When the water with which this succus lapidesceus is mixed, is in part wasted by the Sun and air, it does then deposit it, as being too heavy for it any longer to bear it. And when that is deposited and fallen down, it does by a continued addition and concretion in time amount to a considerable stony moss, &c.

From whence we may note,

1. That by his observation and judgment, the stony substance bred by the water, is nothing but the apposition and fixing of the small stony particles hid in the water one unto another, which is merely aggregation, and so comes to increase the bulk or quantity by continual addition. And thus far, according to this ingenious person and learned chemist, here is nothing at all of transmutation, but that the moss, leaves, &c. become seemingly of a perfect stony nature and hardness.

2. Yet if we look a little more warily, we shall find not only an aggregation of these small stony particles, and an incrustation upon the outside of the moss and leaves, but even that the substance of the moss and leaves, and the small atoms of them are merely petrified as far as our eyes, or the best microscopes can inform us.

And though the thicker and greater pieces of wood, be not in so short a time petrified, as are moss, grass, and the leaves of trees, yet in a longer continuance of time, wood of a considerable bulk, will be totally stonified both in the internal and external parts: So that by this water of the Dropping Well, stones are not only bred by aggregation of small stony particles, nor wood and moss only crusted over with a stony concretion, but also that the moss, leaves and wood, are really changed into a stony substance.

And though the explication of the true manner and way how it is done may be occult, and yet require the study of many observers, and does not belong to our present enquiry, curiously to search forth: Yet thus much is evident, that upon the supposition, that the moss and leaves, &c. and the small parts of them are truly changed into a stony nature, that then the aerial part, or the globuli aetherei (as Cartesius calls them) are by the entry of the stony particles contained in the water, extruded, and so separated.

Whereby two of the particulars are made clear; First, that there is something separated from the thing changed that was in it before, and also that there was something added, as the stony particles, or petrifying steams or atoms that was not there before, and consequently that there must be an alteration of the contexture and position of the particles of the body changed.

But because the stress of the matter lies what transmutation is, which we are searching after, and yet it may be doubted whether or not there be any real transmutation at all, quoad naturam, sed solum modo, quoad nos, as when by a due proportionable commixture of sand and ashes, glass is made, which quod nos, and in relation to our sight is transparent, which

4

neither sand nor ashes are; and yet the sand and ashes in their primitive nature and principles remain as they were, the individual particles of either of them being not changed, as may appear by the reducing them to the same sand and ashes that they were before, which may be made manifest not only by the Alkahest, (only known to adepts) but also by other means that may and can be thrown by expert artists: And also when that silver is dissolved in Aqua fortis, according to our sight it is changed, and the water remains transparent, and the silver may again be separated from it, as is known unto every expert goldsmith, it may very well make us doubt whether there be any real transmutation or not, but what is by addition, diminution, or altering of parts.

And therefore we shall quote some more instances, omitting that of Hector Boelius of the Pond in Ireland, that if a piece of wood be stuck down in it, as much of it as is in the earth or mud, is in the space of a year turned into stone; that part which is in the water, is turned into iron; and that which is above the water does remain wood; so that the same entire piece is stone, iron, and wood; which were a most strange relation, and fit instance if true. But though we have had it affirmed by a learned physician that lived long in Ireland, and that others do maintain that our Irish stones or whetstones are of the same petrified wood, (as the grain or bait would almost persuade) yet because the before cited author is noted to be fabulous, and much suspected in many things and it not proved by later Authorities; therefore, I say, we shall omit it, and so come to some more that the faithful and learned Helmont has noted from authors of better credit, who tell us; For so the hand-glove of Frederick the Emperor was petrified in that one part of it that lay wet in the spring, but the other part being fenced with a seal, remained leather: so that not only herbs, woods, bread, iron, eggs, fishes, birds, and quadrupeds, were by a wonderful metamorphosis petrified; but by the testimony of Ambrosius Pareus, a child at ripe age was cut out of the womb petrified, which his friend told him, that made mathematical instruments, used the back of that petrified child for a whetstone; and more to the same purpose he relates in the same chapter, from whence amongst others he draws these conclusions.

1. That whereas other seeds require that the substrate, or subject matter be reduced into a sequacious, or an obedient liquor, and susceptible of the seed, which they have called the first matter of generation, and do require that also that the figure, and all the comeliness of the precedent concrete be destroyed: yet the petrifying seed, the human figure, being preserved, without any intervening putrefaction, or dissolution of the matter, does petrify the whole through the whole, to wit, as well the bones as the skin.

2. That the petrifying seed does consist alone in a saxeous [petrifying] or stony odour or steam, which is an incorporeal and invisible ferment. We shall not here quarrel with this experienced and learned author, but only note these two things.

1. That whereas he places this petrifying quality in an odour or steam, which he makes invisible and incorporeal, I take him by incorporeal, not to mean merely that the steam is altogether spiritual, as the schoolmen and metaphysicians understand, but that it is so subtle, tenuous and fine, that it is not liable to our sight; and in regard of other more gross bodies, may be called and accounted incorporeal.

2. He plainly holds forth petrifaction, not only in the superficial parts, but that it is totum per totum, as well in the bones as in the skin, not only by incrustation, or adhesion of the stony matter to the external parts, but by a real changing in animals of the bony, sinewy, muscular, and fleshy parts intrinsically and throughout, into a stony substance; to verify which, more authorities may be added to Helmont as that of Pensingius historia infantis, in abdomine inventi, & in duritiem tapideam conversi.

And something of this nature in that accurate and ingenious piece of Mr. Hooke's *Micrography*, as also much of this nature may be seen in Mr. Boyle's Essay of firmness, and in some other places, to which I remit the reader.

Now in all this that the learned Helmont has noted, or the rest, it will appear that this saxeous odour, or seminal ferment, how thin or fine soever it were, is of a bodily nature, and so piercing the body to be changed, whether of animals or other things, as iron, eggs, leather, or the like, it does add some such steams and particles as were not there before, and so augments the quantity or weight, if not both; which was one of the things required to be proved.

Again, by the ingress (which must be by motion) there must of necessity be a cession of another body, which can be nothing but the airy atoms, or ethereal steams, contained before in the porous parts of the body to be changed, which are thereby extruded and separated; which was another thing to be proved.

And as for the third, it necessarily follows, that when a softer body is changed into an harder, or a more fluid body into a firm, the parts are joined more close together; and however all motion in bodies must of necessity make a change of the position, contexture, and order of the minute or smaller parts.

By all which we shall only urge thus much, That this petrification is as strange as that which the philosophers call the transmutation of metals; as may appear by the comparing of their efficient causes, the manner of their operations together, and also of their effects.

To make which plain, we shall here once for all lay down the requisites, all manner of metallic transmutation; and so as, we go, show their accord or disagreement.

And first, in the philosopher's transmutation of metals, they have their subject which they intend to transmute, (to use that common word, though truly the thing that they do, is only to maturate and meliorate) which is some of the more base metals (as they are commonly styled) for as they never mean to change gold into gold, for that would be no change: so in the intent of Nature and its operation, a stone cannot be said to be changed into a stone.

And in this the transmutation of metals, and of changing wood, moss, leaves, animals, iron, and the like, into stone, does agree that they both have a substrate, or subject matter to work upon, and so the one not to be wondered at more than the other: but there are two properties wherein they differ.

1. For firstly, in the petrification wrought by Nature, the things changed are not always contained under the same proxime genus, and the thing working the effect of stonifying is of a lapideous or mineral nature, and (according to common opinion) neither contained within the animal nor vegetable kingdom, and yet are wrought upon by that petrifying agent, when in the philosophic transmutation, the thing changed is under the same proxime genus, with the nature of that which it is changed into, being both of a metallic root and nature, and so is less wonderful than the change made by petrification.

2. Secondly, the things wrought upon by the petrifying agent, are more remote from that stony nature into which they are changed, whether they be animals or vegetables, as having had no previous preparation, to fit them for the susception of the operation of that petrifying power: whereas in the transmutation of metals, the metal to be changed is to be made as clean as Art is able to perform, according to that true and certain rule of our countryman Ripley, who said:

"For who that joins not the Elixirs, with bodies made clean,
He woteth not sykerly what Projection does mean."

Secondly, and as the agent in the change wrought by petrification, is (according to the doctrine of Helmont) a petrific seed, consisting only in a saxeous odour, or invisible ferment: So the agent in metallic transmutation, is a seed of an aurific or argentific nature, for it is known to all that are masters, that the Elixir or small part of that which they call the Philosophers Stone, or Tincture, has a seminal power, able to produce its like, according as it was specificated by fermentation.

Which is sufficiently confirmed by that faithful description that the experienced Pole has given us, who said, "*Lapis Philosophorum, seu Tinctura nihil aliud est quam aurum in supremum digestum, nam aurum vulgi est sicuni herba sine semine, quando maturescit producit semen, sic aurum quando maturescit, dat semen sive Tineturam."*

["The Philosopher's Stone, or Tincture, is nothing but gold sorted into the highest, for the gold of the common people is like a plant without seed, when it ripens it produces seed, so gold when it ripens, it gives seed or tincture."]

And again, "Aurum potest dare fructum & semen, in quo se multiplicat industria sagacis artificis, qui scit naturam promovere, sed si absque natura, id velit tentare, errabit."

["Gold can give fruit and seed, in which the energy of a skilful craftsman, who knows how to promote nature, multiplies; but if he wants to try this without nature, he will err."]

To which does agree that often quoted saying of Augurellus.

"Hordea cui cordi demum ferit hordea : ne tu
Nunc aliunde pares auri primordia: in auro
Semina sunt auri, quamvis abstrusa recedans
Longius, et multo nobis quaerenda labore."

[Finally, the barley strikes whose heart the barley strikes. Lest you should now compare the origin of gold from other things, in gold are the seeds of gold, although they are obscured by receeding farther, and we have much to seek with labour.]

So that as they agree in having a subject matter to work upon, and in their agents, that they are both of a seminal power; so there is no cause to account the one strange or impossible, and the other not, except it be by reason that petrifying is more common, and the change of metals, but seldom or rarely seen, which though it might stagger vulgar brains, yet can be of no moment to a learned and considerate mind.

Thirdly, they agree in the manner of their operations, for in the act of petrification, there is the saxeous odour or seminal ferment added to the thing changed, that was not there before, and the airy steams or particles are extruded forth of the body changed that were there before, so that the position and contexture [weaving together] of the small particles of the body changed are thereby altered and changed.

9

So in the transmutation of metals, there is added some small part of the Philosophers' Tincture unto the metal (as suppose it lead, or quicksilver) that is to be changed: And also there is something that is separated from the body changed that was in it before, as less or more of that which some call the external, separable and combustible sulphur; but Trevisan calls it scoria dross, which being indeed of an heterogeneous and differing nature from the homogeneous mercury, did make its small parts that were homogeneous one to another, that they could not so nearly be joined per minima, which by the ingression of the small quantity of the tincture are extruded and separated.

But to illustrate this, we shall give the unquestionable testimony of Helmont's experience, who said in his *Arbor vitae* p793, *Cogor credere lapidem aurificum, & agentificum esse, quia distinctis vicibus manu meâ unius grani pulveris super aliquot mille grana argenti vivi ferventis, projectionem feci, &c.*

[I am impelled to believe that the stone is golden and argentive, because I made a projection with my hand on several occasions of one grain of powder upon several thousand grains of boiling quicksilver.].

And that a great multitude standing by, the matter, with the tickling admiration of them all (it seems himself not excepted) did succeed in the fire, as the books of that Art do promise. And that he that first gave it him (so that it seems he had either given a second time, or more, or else he had made it himself, because he had of divers proportions) was a stranger, and but a friend of one nights acquaintance, and had at the least so much as was sufficient to change two hundred thousand pounds into Gold. And that he gave him half (a grain he calls the sixtieth part of a dram) and from thence nine ounces and a quarter of quicksilver, were transmuted, which was a high proportion and noble exaltation.

Again, that he had once given him (which differs from the other) the fourth part of a grain, which fourth part of a grain being wrapped in paper, he projected upon eight ounces of hot quicksilver in a crucible. And forthwith the quicksilver with a

10

certain noise, staid from the flux, and settled like a yellow lump or morsel, and after being melted, with the blast of the bellows, there was found eight ounces of pure gold, wanting eleven grains. Therefore that one grain of that powder had transmuted into the best gold, of quicksilver equal parts to itself 19186: which was a most noble multiplication exceeding the former.

Again, he confesses that sometimes, at divers times he had handled it in his hands, and with his eyes seen the real transmutation of common vendible quicksilver, in proportion exceeding in weight the gold-making powder some thousands of times; and that it was of colour like saffron in its powder, but very ponderous. and shining like beaten glass, when it is less accurately made into powder, and that once the fourth part of a grain was given him of it. And this he inclosed in wax, lest in throwing it into the crucible, it might be dispersed by the smoke; which he projected upon a pound of hot quicksilver, newly bought, and put into a crucible. And forthwith the quicksilver, with a little murmuring noise, staid from the flux, and settled to the bottom like a lump. And that the heat of the quicksilver was but so much as might hinder melted lead from recongealing. Then, by and by, the fire being increased under the blast of the bellows, the metal was melted, and the melting pot being turned upwards, he found it to weigh eight ounces of most pure gold.

And a computation being made, a grain of that powder did convert 19200 grains of impure and volatile metal, that may be put away with the fire, into pure gold; only in this there is required a moderate fire of glowing, or burning coals. And this is a higher multiplication than the former.

From all of this we may note,

1. That these were three several sorts of powders, differing from, or exceeding one another in nobility and virtue.

2. It is probable that in the last mentioned projection, he was not punctually acquainted with the quantity upon which he was to project it; otherwise he would have cast it upon less than one pound; which produced but eight ounces, the other flown being flown, or otherwise wasted in the fire.

3. From hence we must note, that in projection the metal to be changed, is to be in flux and open, that the gold-making powder may the more easily have ingression, and penetrate into the smallest parts of the metal to be changed; for Paracelsus tells us, (*Rerum Naturae* I.7. p. 97), that as water being hardened by cold into ice, will not receive the tincture of saffron in powder cast upon it; but when melted into water, easily will: so the metal to be changed must be in flux motion, and opened by the fire, otherwise the tincture cannot have ingression nor spread itself, and where there is no ingression there can be no transmutation.

Yet here Helmont tells, that it need but be easily hot, and not violently to any great degree, but as much as may keep melted lead from recongealing. And this previous artificial help, besides the cleansing of the metal to be changed as much as Art can perform, is requisite in metallic transmutation, though in that wrought by Nature in vegetables, or animals, in petrifying of them, there is no such precedent preparation, nor adjuvant cause, as external heat or fire, but the petrifying steams, or seminal odour, does effect the thing without such helping concomitants, so that (if duly considered) the work of Nature, without the assistance of Art, in petrifying of vegetables, and animals, is more strange and wonderful than the transmutation of metals.

4. We may note, that Nature in changing vegetables or animals into stone, does often work pedetentim [step by step], and by degrees, as also sometimes subitaneously [hastily] and quickly, as may appear by that story of Helmont, which he thus relates. About the year 1320, between Russia and Tartaria, in the altitude of 64 degrees, not far from the pond or fen called Kitaya, it is read, that a hoard of the people called Baschirde with their whole herd of cattle, their wagons and carriages were altogether transmuted into rocks or stones.

And that yet to this day the men, the camels, the horses, the flock or herd of cattle, and every other kind of thing. that did accompany the wagons or carriages, do yet standby a horrible spectacle, in the daylight turned into stone, and that this was done in one night, without any preceding putrefaction. The like story (if my memory fail not, for I have not the author by me) is in Ola Magua, an author of good credit and reputation, and the like may be found confirmed by some other writers. Which (if true, and no miracle) shows that this act of petrifying of vegetables and animals is sometimes quick and subitaneous, as of one night, only that change of metals is done in a far less time, and therefore may well be said to be an acceleration of the work of Nature by the help of Art.

5. It may very well be believed, that in the changing of vegetables or animals into stone, that the thing changed is of more ponderosity, and for the most part of greater bulk than the thing was of before it was so petrified and changed. For so we have found in all our trials of wood, moss, leaves, and the like, stonified by the Dropping Well near Knaresborough, because that is done by incrustation, but whether it happens to be so in all other sorts of petrification (for doubtless there are more ways than one) our experience cannot determine, but must leave it to the trial and examination of others.

But in metallic transmutation, if the exact degree of the virtue of the powder transmuting be known, and so be projected upon a just and due proportion, the ponderosity will not much differ from what the metal changed was of before, as appears by that experiment of Helmont's, where he projected one fourth part of a grain of the gold-making Powder, upon eight ounces of hot quicksilver, and it produced eight ounces of pure gold, wanting eleven grains, so that here was no great difference in the weight. For reckoning that the eight ounces of quicksilver, had the fourth part of a grain added to them, and when changed into pure gold, had but lost ten grains and three quarters of a grain, which must be that either the quicksilver had in it so much of combustible sulphur (as Helmont in a certain place of his writings confesses that all common quicksilver has in it less or more of combustible and separable sulphur) that was separated or wasted away in the fire: or that so much of the homogeneous body of the quicksilver did evaporate as being made too hot, and either of these ways it might have been, though the first is most certain. that all imperfect metals have less or more separable and combustible sulphur, which in projection is separated and wasted.

But howsoever that there be little difference of weight in the metallic body changed from what it was before, yet it always becomes less in bulk, and possesses lesser room, or place, as appears by this of Helmont, that the quicksilver settled with a certain noise to the bottom of the crucible, and so became of less bulk, and possessed less room.

And that this is, and must be so in all metallic transmutations, is most clear, not only from the authority of the Adepts, but from their convincing reasons, showing that in their transmutation, there is a radical solution and penetration of all the small parts or atoms of the metal to be changed, by the subtle penetrability and ingression of their so much purified and exalted tincture, and thereby all things in it whatsoever that are of an heterogeneous nature, are separated and extruded, and the (homogeneous particles joined together per minima as much as Nature can admit of, and so must needs be of less bulk, and possess less room or place, which is manifest in gold, that is one of the heaviest bodies in the same bulk that Nature does produce, as being most dense, containing most of matter, and having its particles most closely joined together, that there are few interstitials or spaces for the air or aether to enter or possess, which is manifest in its extension under the hammer, whereby it will be foliated farther, and be thinner than any other metal whatsoever; and so a baser metal changed into gold, must of necessity possess less room, and be of less bulk.

6. And that we may come a little nearer to manifest this great work of the transmutation of metals, we may consider, that though in petrification by the seminal ordour or saxeous ferment, it works upon most bodies as it finds them, either more susceptible or more apt to resist, which might render its operation and effects more difficult and strange. But here the matter is rendered more feasible and facile, not only by a previous cleansing of the metal to be changed from its heterogeneous parts, and gently opening of its body by fusion in

an easy fire; but also our Learned Countryman Roger Bacon (*Speculum Alchymicum* ch7, p269) does show plainly that we having nearer metals unto the more noble, are excused from the more remote: for seeing that Saturn, Jupiter and Mercury are more near than Venus or Mars, we were foolish to take the latter, and to leave the former.

7. The ancient philosophers that were masters of this great secret of transmutation, and knew it by experience, and had seen it with their eyes, took little care of framing methodical definitions or descriptions of it, as little valuing such trifles and niceties, but contented themselves with the true understanding of it; and yet to their disciples which they termed the Sons of Art, they gave sufficient hints of the way and manner of it, but still as veiled and obscured.

But I find that Paracelsus (however condemned of many for his too dark writing) to have said more of transmutation in general, than the most of those that went before him; some of which we shall here recite, where he says (*Liber rerum naturae*, 7, p97) thus: "If we shall write of the Transmutation. of all natural things, it is equal and necessary, that before all things we first show what transmutation is. Secondly, what are the degrees to come unto it. Thirdly, by what means, and after what manner it is done."

Therefore transmutation is when a thing loses its form, and is so altered, that it is altogether unlike its former substance and form, but assumes another form, another essence, another colour, another virtue, another nature or propriety; as if a metal be made glass or stone, if a stone be made a burnt coal, if wood be made a coal, clay be made a stone or brick, a skin be made glue, cloth be made paper, and many such like.

Now though this be far from a logical definition, as written by one that is generally believed to be no friend to logic; yet is it no bad description of transmutation in general, and may well stand uncondemned, unless by those that can produce a better: for if the things that he does instance in to be changed be duly considered, the most of them have incidents in the way and mode

15

of their transmutation, that are as difficult to explicate and declare as the principal things in metallic transmutation. Is it not hard to open the true causes how Antimony, that is a metallic body, is per se (which every common Chemist can perform) brought into glass, which is a transparent body, the matter considered, will not be found so easy? And so (if we had leisure) might be said of some of the rest.

8. And that we may more plainly understand the manner of this metallic transmutation, let us a little consider the virtues and properties which they ascribe to their tincture when perfected, because by it the operation is performed: for if the nature of the agent be well known, the effects that it works upon the patient may be the better perceived; and they are thus enumerated and described by that ingenious and experienced person Johannes Spagnetus, who said, "There are five proper and necessary qualities in the perfect Elixir, that it be fusile [able to be melted], permanent, penetrating, colouring, and multiplying; it borrows its tincture and fixation from the leaven, its penetration from the sulphur, its fusion from argentvive, which is the medium of conjoining tinctures, to wit, of the ferment and sulphur; and its multiplicative virtue from the spirit infused into the quintessence."

From whence we may gather not only its virtue and energy, but in some measure its manner of operation.

1. For, first, we are to note that all that are properly called metals, that are to be changed, are fusile, and apt to be melted, and flow with the force of fire, though some more easily than others; and if the tincture which is the efficient changing, were not of a fusile and flowing nature, it could never mix or conjoin itself with the metal to be changed; for where there is no ingression, there can be no mutation.

2. That otherwise it could cause no transmutation, for nihil dat quod non habet [he gives nothing that he does not have]; and by these two properties all heedful and considerate persons may easily conjecture, from what root it must needs originally arise, and so may truly know the first matter.

3. It is of a most penetrating nature: for if it were not so, the small and homogeneous atoms of the metal to be changed could not be pierced, and thereby to be so ordered that they may be joined per minima, and united together, and thereby to extrude whatsoever is heterogeneous in the metal to be changed.

4. It has also the property of colouring, being indeed the sulphur, or fire of Nature, from whence all colours do arise; and mixing itself with the metallic mercury of the body, or metal to be changed, which radically in all metals is one and the same, it becomes one with it as arising from the same root; and so by the help of Art accelerates the work of Nature, and does that in a short time, that Nature cannot perform in many hundred of years, as says the learned Philosopher in these words: *Et haec est auri forma, summum & optimum, quod ad metallicam naturam spectat. Si itaque pura istius modi forma, quae per artem, mediante natura, praeparari potest, imperfectis Metallis addatur, tunc impurum imperfectorum Metallorum superatur. Non enim impurum, sed pura materia illi est similis: Prima siquidem est forma ad quam materia ista facta fuit. Idcirco par cum pari tempore incomprehensibili conjungitur, impurum separant, quasi dicant: An tu venisti, quod meum est, & quod ad me spectat?*

[And this is the form of gold, the highest and best, as regards the metallic nature. If, therefore, a pure form of this kind, which can be prepared by art, through the mediation of Nature, is added to imperfect metals, then the impurity of imperfect metals is overcome. For it is not impure, but pure matter that is similar to it: indeed, it is the first form to which that matter was made. For this reason the equal is joined with the equal at an incomprehensible time, they separate the impure, as if to say: Have you come, that which is mine, and that which pertains to me?]

5. It has a power to multiply the virtue, but not the quantity; and having these rare qualities, it is no such wonder that it should work such effects upon the more imperfect metallic bodies.

9. And that we may more clearly apprehend the nature of this transmutation, we must consider some of their maxims; which though by many slighted, yet do they hold forth the certain and absolute truth:

1. As first, that of Bacon, (*Speculum alchemiae*, Ch3, p269) which they all allow of as the basis of all philosophic verity, which is this, speaking of sulphur or Nature's fire, and mercury's natural or radical moisture, he said, *Sed ex praedictis duobus fiant Metalla cuncta, & nibil eis adhaeret, nec eis conjungitur, nec ea transmutat, nisi quod ex illis est.* [But from the two aforesaid, all metals are made, and the mist adheres to them, and is not joined to them, nor does it change them, except what is of them.] Which is a golden sentence, containing both truth and plainness to those that will rightly consider, and understand it.

2. Another is this of the same Author (*Speculum alchemiae*, Ch2, p258): *Sed dico quod natura semper proposuit, & contendit ad perfectionem auri. Sed accidentia diversa supervenientia transformavit metalla, sicut in multis invenitur Philosophorum libris satis aperte.* [But I say that Nature has always proposed, and strives for the perfection of gold. But the various accidents that occurred transformed the metals, as we find quite openly in many books of the philosophers.

3. A third is this (*Musaeum Hermeticum* Ch2 p41): *Est itaque omnibus in Metallis verus Mercurius, rectumque Sulphur, aeque tam in imperfectis, quam perfectis Metallis: Saltim contaminatus, & impurus factus est in imperfectis Metallis, & quae sola perfecta maturatione destituuntur. Et ex iisdem causis ad aurum, argentumque redigi possunt, h.e. ut ab aurea, vel argentea natura, quae in illis est, separetur impuritas, qua cum inquinata fuerant, & forma auri, vel argenti iisdem ingeratur.*

[Therefore in all metals there is true mercury, and a true sulphur, and equally in imperfect as in perfect metals. And from the same causes they can be reduced to gold and silver, i.e. so that the impurity with which they were defiled may be separated from the golden or silver nature which is in them, and the form of gold or silver may be infused into them.]

4. A fourth is, That all metals are in suo interiori, gold, silver, and mercury, and that metallic mercury can no ways be destroyed, or otherwise the art of transmutation were utterly false, which is certain, true, and most true.

10. From all this we may plainly gather what the transmutation of metals is, and how it is wrought: So that if metals be in their root all of one mercurial and homogeneous nature, and that there be perfect sulphur and mercury equally as well in the imperfect as perfect metals, then must their transmutation be easy. For then the heterogeneous matter, or combustible sulphur, scoria, or dross, being removed, and some of the tincture added, the parts are most closely joined, and so united per minima, and tinged, by which means they are maturated in a short time by the help of Art, that Nature could not perform in many years.

So that all metallic Mercury wants nothing of the degrees and nature of gold; but removing of its heterogeneous parts, and the adding something more of the fire of Nature, and then it becomes most dense, and to have all the requisites that are necessary to gold. Agreeable to what we say here, is the opinion of an ingenious person, who in the *Philosophic Transactions* No41, p813, says thus: "To conclude, I shall presume to give you some of my thoughts concerning the so much discoursed of transmutation of metals; concerning which I am of the opinion, that the change is erroneously apprehended by many, imagining that the whole imperfect metal is totally transformed into the more perfect by the substance mixed with it; where as the mixture added to the melted metal, joins itself (as I conceive) to those parts, which being homogeneal, symbolise together with the nature of the more perfect, whereby the pure metalline parts are separated from the other heterogeneal impure sulphurs; which, together with other causes, did hinder Nature in the mine from concocting that substance into the perfecter metal.

A second instance that we shall give, is, that divers vitriolate waters do change iron put into them into copper, which Helmont does deny to be any transmutation, and says thus (*De Spadan.* Sent. 1 Paradox, 3, p692): But that vitriol-bearing juice is thought to change iron into copper, the mine-men themselves not acknowledging the delusion, because that the succeeding atoms of the copper do fill up the place of the iron that was wasted; neither regarding that as copper does render or make silver dissolved in aqua fortis, that otherwise was invisible, to appear to the view, and be corporeal.

So that it is the propriety of iron dissolved in the vitriol to manifest the copper by drawing it to itself, and together in the same act, that the iron itself is dissolved, and does vanish in the fountain. My witnesses (he says) are the fountains themselves; because verily the vitriolate waters are far more poor in copper than they were before the iron dissolved in them, and the copper thereby recovered from them. Therefore to wit verily out of the very fountain (where it is often continued, the flux of new copper does fail in the pit or spring) the putatitious transmutation of iron does otherwise not happen.

The manner of doing of which in the mines of Hungary, called Herrengrundt, Athanasius Kircher does thus describe (*Mundus Subterrraneus*, 1, 10, Sect 4, Ch 10 p.123, 24) "They take rusty iron that is unprofitable, as the remainder of various and old instruments used in houses, and being put into the furnace and made hot, they are upon the anvil beaten forth into most thin plates. This being done, they put these plates into the bottom of vitriolate water, which does flow in the most deep pits of the mines; and being put there, they leave them for certain months. And the due time ended, they come to the pit, and find the plates to be gone (or changed) into a yellowish stuff, like unto a soft plaster, and these exposed to the air and winds, is hardened into copper of the best account."

And it is so used at Neosel in Hungary: Therefore it is questioned whether this be a true transmutation of iron into copper, or not. But I say that here true transmutation is not at all given, seeing that all the whole iron is not changed into the substance of the copper, but by accident only, I do explain myself. For seeing that in vitriol infinite copperish corpuscles do inexist [exist inherently], and as those have the greatest sympathy with iron, so that also it comes to pass, that forthwith they flow unto the iron, and do most intimately insinuate themselves into its pores; but seeing that they abound with spirits of great acrimony, from hence being insinuated into the iron, forthwith they begin to corrode it, so far, that all the fatness of the iron being consumed, the irony substance being dissolved, does pass into dust, or a rusty powder, the vitriolate corpuscules substituting themselves into the place of the iron being consumed, and the native particles both of the iron and vitriolate water are conglutinated into one mass, which first truly is soft within the water, but being exposed unto the more free air, the wind and beams of the sun are indurated into perfect copper, and by this means it is made the same thing that it was before: before verily by the dispersion of its corpuscles in the waters, now by the union of the same attracted from the iron.

But if here were given a true transmutation, nothing of the iron should remain after. But experience teaches, that so much of the irony rust does remain, almost as much as the irony mass did weigh before. And after he shows an experiment, by a rod or thread of iron put into some of this vitriolate water sent him forth of Hungary, in this order: I (he says) put an iron thread into a vial full of this water, which in the space of three days was all consumed, a certain soft matter remaining in the bottom, which separated from the dross, did yield pure copper; but the dross remaining, did almost come to the weight of the thread of iron; so that from hence no man need further doubt of this matter. Thus far the experience and opinions of these two learned persons touching thus kind of change, which they will not allow to be a true transmutation, from whence we shall move some considerable observations, and submit them to the judgment of those that have learning and leisure to examine the pertinency and validity of them.

21

1. And first, it this (in their sense) be not a true mutation, yet of necessity it is an apparent one: for the iron not only to sense had in it the requisites that are accounted proper to that metal, but also really had that which all account the properties of that metal, as to indure ignition, extension by the hammer, and fabrication into instruments, which by being brought into copper, has not only a more glorious colour than that of iron, but will indure ignition even to fusion, and that more easily than any iron, and is become more extensible than iron, and admits of more easy fabrication into instruments. So that this change, (of what sort soever it be taken to be) is a meliorating of the thing, a graduating and exalting of it both in intrinsic and extrinsic virtue, the metallic root or nature still remaining. So when the philosophers mention the transmutation of metals, as the changing of lead or quicksilver into gold or silver, they do but understand a bettering, exalting, and graduating of them, the metallic root still remaining: so that there is no such great difference as many ignorantly do conceive and imagine.

2. Secondly, if they mean (as they seem to hold forth) that no transmutation is true, but where all the atoms and corpuscles of the body to be changed, are every and all of them transmuted, without separating of any of them, or adding any thing unto them, then we must say, that (as far as we either know or understand) few such transmutations will be found in rerum natura, brought to pass either by Nature or Art. And for the metallic change that the philosophers speak of, they never held that all the atoms or particles of lead and mercury are transmuted into Sol or Luna, but that the homogeneous parts only are, and the heterogeneous parts separated by the addition of some part of their noble stone, which is not much differing from this mutation of iron into copper.

3. We may consider the manner how this change is done, and that is by taking it for granted, that in the iron before it be changed, there are store of corpuscles of copper, as also in the vitriolate water, and the water by its acrimony corroding the iron, and thereby separating the atoms of the iron, those of the nature

of copper residing in the said water, do substitute themselves in the place of the atoms of iron, being separated; and so being atoms of a congruous figure, size, and other properties, do easily couple themselves together, as being homogeneal, and refusing others as of a disagreeing nature. So the masters do hold that their stone when exalted and prepared to the red, is aurum intensum, exuberatum & animatum, as being indeed brought and wrought from a golden seed and that the homogeneous mercury of all metals, is in suo interiari of a golden nature, these two easily unite most closely together, and refuse union with any heterogeneous body, and so the manner of both these changes are alike.

4. It would be worth labour to examine the certainty, whether all iron, or the ore from whence it is drawn, have something of the corpuscles of copper in it, and (if possible) in what proportion: That thereby it may be considered whether the atoms of copper be in the iron, and the atoms of iron in the copper, by accidental commixture, or that they come to be so by progressive generation.

And then it may be considered, that where there is particles of copper and iron mixed in one body, which seems to be iron, and to which we give that denomination, be when it will, or its ore is found so, in its ascension or descension, as the mineralists speak, that is, whether in continuance of time more copper would increase and grow in it, or that in length of time the copper atoms would decay or grow into iron? A query that may be necessary for all lovers of mineral knowledge.

5. There is a passage in that profound, though dark piece, written by Paracelsus, which is commonly called *Caelum Philosophorum,* or *Liber Vexationum,* though some of great judgment call it *Liber Fixationum,* which here may well be considered of, and that is this (*Can.* 1, p121); *Omnia sunt in omnibus occultata. Unum ex ipsis omnibus est occultator ecrum, & corporeum vas, extrinsecum, visibile, & mobile.* [Everything is hidden in everything. One of them all is the occult concealer, and a corporeal vessel, external, visible, and mobile.]

This hint with divers others, in the obscure and enigmatical writing, though not regarded by many, that are so idle and lazy, that they will not take pains to break the hard shell, thereby to gain the precious kernel, not minding that *rosa non nascitur sine spinis* [a rose is not born without thorns], and that *Dii sua bona laboribus vendunt* [the gods sell their goods for labor], do sufficiently show, that the nature of metals is not yet perfectly understood. And to me by this he seems to intimate that all metals are hid in all metals, and that one is the hider of them.

And therefore the question pertinent to this case, will be, whether the iron does hide the copper, or the copper the iron, and so of other metals; which we shall nor decide, but leave it to the judgment and trial of others.

A third instance that we shall give, is in an artificial transmutation (if we may call it so) and that is of quicksilver, which is a fluid, open and volatile metallic body, and yet is and may be by art brought into a firm, close and fixed body, as Helmont declares thus at large. *De Febribus*, Ch. 14, p52). There is also the purgation *Duceltatesson, quae Podagram non minus, quam febres radicitus curat. Ejusque arcanum corallinum vocatur, quod paratur ex essentia auri Horizontalis, hoc modo. A Mercurio vulgo venali, abstrahe liquorem Alkahest, cujus meminit 2. de viribus membrorum, c. de hepate. Quod sit unius horae quadrante. Nam, inquit Raymundus, astantibus amicis & praesente Rege, coagulavi argentum vivum, & nemo prater Regem, scivit modum. In quâ coagulatione istud est singulare. Quod liquor Alkahest idem numero, pondere & activitate tantum valet millesima actione, quantum primâ. Quia agit sine reactione patientis. Mercurio igitur sic coagulato, absque ullam coagulantis remanentia, fac inde pulverem minutum, & destilla ab illo quintes aquam ab albuminibus onorum destillatam, atque Sulphur Mercurii, quod per sui praefatam coagulationem foras deductum est, fiet rubicundum instar coralli: & quanquam foeteat aqua albuminum, tamen iste pulvis dulcis est, sixus, ferent omnem follium ignem, nec perit in plumbi examine.*

Spoliatur tamen virtute medicâ, dum in album metallum redutitur.

[Duceltatesson, a Paracelsian purgative cure for fever, which cures gout no less than root fevers. And his secret is called coralline, which is prepared from the essence of horizontal gold, in this way. From Mercury commonly sold, extract the liquid Alkahest, of which he holds in memory for the virile member and for the liver. That is a quarter of an hour. For, says Raymond, in the presence of my friends and the presence of the King, I coagulated the living silver, and no one but the King knew the method. In what coagulation this is singular. That the liquid Alkahest is the same in number, weight, and activity is only worth a thousandth action, as much as the first. Because it acts without the reaction of the patient. Take mercury, therefore, thus coagulated, without any coagulant remaining, make a fine powder from it, and pour from it five times the water distilled from the albumen of the clouds, and the sulfur of Mercury, which has been brought out by its aforesaid coagulation, will become red like coral: and although the water of the albumen stinks, yet this one is of sweet dust, fixed, bearing the fire of all follies, and does not perish in the examination of lead. It is, however, stripped of its medicinal power, while it is reduced to white metal.

A relation of this notable experiment and most strange mutation may also be found in the *Theoria* of Raymund Lully the 87th chapter, which the learned reader may consult and consider of. But from hence we shall observe these few things.

1. That this seems to be a more strange mutation than any other we can meet with, for by this the common mercury, an open, fluid, tremulous and volatile body, is made a shut, firm, settled and fixed body, even to abide all the fire of the bellows, and not to perish in the trial of lead, which is all that silver, will endure.

2. Here is nothing at all added unto it, but which is again wholly separated from it, for the Alkahest is drawn of the same in number, weight and activity, leaving not the least atom

25

remaining with the Mercury: Whereas in the transmutation of metals by the Elixir, the part of the powder projected does remain inseparably with the changed metal, so that of the two, this act of the liquor Alkahest upon the mercury is more strange than that of the Elixir upon another metal.

3. They agree in this, that in the transmutation of metals by the Elixir, the extraneous sulphur, and heterogeneous parts (which in quantity less or more are in all metals) are removed and separated, and so in the fixing of the mercury the extraneous sulphur, is extroverted and turned to the outside, by the operation of the Alkahest, which sulphur contains in it the medical virtue, which by melting down is wasted, consumed, or separated, and so the change in both is made by separating something from the changed body, that was in it before.

4. By this it is manifest that in both these mutations, the mercury by the Alkahest, and some other metal by the Elixir, both aster the change become of less weight than they were before, according to the quantity of the heterogeneous parts separated from them.

5. Lastly, the mercury is fixed by having the extraneous sulphur thrust from betwixt the homogeneous atoms of the mercury, and thereby they become more closely united per minima, which is the cause, or rather the fixation itself: and the transmutation of imperfect metals is not only performed by the extrusion and separation of their combustible sulphur, whereby their parts may lie more closely together, but also by the perfect union of the powder projected, with the mercury of the metal changed, being both of one radical nature, and of a symbolising and homogeneous quality and condition.

Alchemical Translations Series

35. First Book of Distillation - Della Porta
36. The Philosophical Parergon - Nollius
37. War of the Knights - Limojon
38. Dialogue - Aegidius de Vadis
39. Donum Dei - Samuel Baruch
40. Banquet of the Sages
41. Transformation of the Metals - Denis Zachaire
42. Allegory - Eirenaeus Philalethes
43. The Memorial of Alchemy - Pierre Vicot
44. A Philosopher and a Peasant discuss Alchemy
45. Transmutatory Alchemy - Timothy Willis
46. Light out of Chaos - Louis Grassot
47. The Play of Children and the Work of Women
48. The Secret - Jodocus Grever
49. The Metamorphosis of the planets - Monte-Snyders
50. A Philosophical Riddle - Birkholz / Adamah Booz
51. The guide to the chemical heaven - Jacob Toll
52. Mercury's Caducean Rod - William Yworth
53. Centrum Naturae Concentratum - Ali Puli
54. The Fate of the Alchemists
55. A Cabalistic Fable - Monte Hermetis
56. The Philosophical Bird-Catcher
57. Truth of the Philosophers' Stone asserted
58. Chrysopoiea
59. Twelve Royal Palaces of Hermetic Wisdom - Fictuld
60. The Mystical Cabbala of Nature - Fictuld
61. The Pilot of the Living Wave
62. Philosophia maturata
63. Chaos - Fictuld
64. The Open Ark
65. Steganographic Allegory - Beroalde de Verville
66. The Aphorisms of Geber
67. The Secret Fire - Glauber
68. Anthroposophia Theomagica - Vaughan
69. Allegorical Dream - Fictuld
70. More - Ezekiel's Vision of the Mercava

71. Chymist's Key - Nollius
72. The Fama Mystica
73. The Crowning of Nature engravings
74. Cabala Verior
75. Zosimos
76. Three short alchemical texts
77. The Three fires of the Sophi
78. Rosicrucian Preface - Sperber
79. Of the Elements and the Quintessence - Drebbel
80. Treatise on the Philosophers' Egg - Bernard of Treviso
81. Ariadne's Thread
82. The Philosophers' Stone - Vauquelin des Yveteaux
83. The Inferior Astronomy
84. The Tomb of Poverty - Henri d'Atremont
85. Compendium Hermeticum
86. The Great Work Unveiled
87. Tinctures of the Seven Metals - Basil Valentine
88. Letter from Hephaestion to Alexander the Great
89. The True Hermes
90. Nature Uncovered
91. The Chemical Truthsayer
92. The Blood of Nature - Brummet
93. An apologetic treatise - Robert Fludd
94. The Key to the Great Work - Sancelrian Tourangeau
95. The Mytho-Physico-Cabalo-Hermetic Concordance
96. The Reign of Saturn - Huginis a Barma
97. Salt, Light and the Spirit of the World - Nuysement
98. A Philosophical Letter - Philovite
99. Writings of Cleopatra the Alchemist and Maria Prophetessa
100. A Rosicrucian Colloquium
101. The Golden Treatise of Xamolxides
102. The Golden Rose
103. Explanation of the Emerald Tablet of Hermes - Garland
104. Mercury Revived - Samuel Norton
105. The Compass of the Wise - Birkholz
106. The Silence after the Clamour - Michael Maier

107. The Golden Mirror of Outer and Inner Vision
108. A Hermetical Banquet
109. Addresses to the Gold- and Rosy Crucians - Ecker
110. Chrysopoeia - Augurel
111. The True and Perfect Preparation - Samuel Richter
112. The Hellish Goddess Proserpina - Rudolph Glauber
113. The Powder of Projection - D.L.B. Lord of la Borde
114. Gold Unclothed - Johann Christian Orschall
115. The Divine Arcana
116. Three Curious Alchemical Writings
117. Theory and Practice of the Gold and Silver Trees
118. The Philosophical Water
119. A Treatise on Metals and Alchemy - Bernard Palissy
120. Chemical Essays - Karl von Eckartshausen
121. The Aurora - Henri de Lintaut
122. Of Hyle, the Universal Prima Materia - Khunrath
123. The Four Amphitheatre Engravings - Khunrath
124. The Fire of the Magi - Khunrath
125. The princely and monarchical Roses of Jericho - Fictuld
126. The Oraculum Manuscript
127. Kings of Scheschian
128. The Theoricus or Second Degree of the Rosicrucians
129. Nine treatises on Goldmaking - Stephanos
130. The Philosophers' Stone - Athanasius Kircher
131. Aelia Laelia Crispis - Nicolas Barnaud
132. Sphynx Rosacea - Christophorus Nigrinus
133. The Theory of the Divine Art of Alchemy - Mylius
134. Summum Bonum - Robert Fludd
135. The Philorcium of George Ripley
136. The Hieroglyphics of the Egyptians - Michael Maier
137. Hermes - Emblems of the Twelve Nations - Michael Maier
138. Maria the Jewess - Emblems of the Twelve Nations - Maier
139. Light emerging by itself from the darkness
140. The portable laboratory - Johann Joachim Becher
141. The Hieroglyphics of the Greeks - Michael Maier
142. On the Creatures of the Aethereal Heaven - Robert Fludd